Gleanings

Laurie Michaelis

A *Friend* Publication

All rights reserved
Permission to quote extracts or to
reproduce articles from this book in any form
must be obtained from the publisher

Copyright: Laurie Michaelis 2017
The Friend Publications Limited

Published by:
The Friend Publications Limited
173 Euston Road, London NW1 2BJ

Charity registration number: 211649
Cover copyright: Ivan Mikhaylov (yellowj) / 123RF Stock Photo

Printed and bound by:
Henry Ling Limited, The Dorset Press, Dorchester DT1 1HD
Set in Baskerville type

ISBN: 978-0-9954757-1-7

Printed on 100% recycled paper manufactured
from responsible and renewable sources

Contents

Preface   Ian Kirk-Smith   5

Introduction   Laurie Michaelis   7

Gleanings

The holy mountain	11
Deep nonviolence	13
Living with difference	15
Call to adventure	17
A simple lifestyle freely chosen?	19
Not a notion but a way	21
Transformational listening	23
Being a transformational community	25
People and place	27
Losing liberty?	29
Making a connection	31
Higher power	33
Seeking unity	35

Other pieces

Answering that of God in everyone	39
Joining the movement for a new society?	41
Seeking unity in a way forward together	43
Darkness and new life	45
Back to the land?	47
Living in the End Times?	51
Loving the Spirit of the Age	53

Preface

The word 'sustainability' was highlighted in 2011 at Yearly Meeting Gathering in Canterbury when Friends agreed to what became known as 'the Canterbury Commitment'.

For Laurie Michaelis and other Friends this was the culmination of a long journey and the beginning of another. They had worked for many years to encourage a concern for the environment within Britain Yearly Meeting and an engagement with the challenge of climate change.

This book contains two sections. The first, 'Gleanings', is a series of articles commissioned for *the Friend*. They are personal reflections on important themes and subjects and show the range and depth of a gifted writer whose reverence for the Earth, sensitive insight, respect for the idea of stewardship, and deep spirituality radiate from every page. The second section contains a selection of miscellaneous pieces.

Thanks are due to the trustees of The Friend Publications Limited, particularly Geoffrey Durham, and to Elinor Smallman, production and office manager of *the Friend*, for her work on the book.

Ian Kirk-Smith
Editor, the Friend

About the author

Laurie Michaelis knew as a teenager in the 1970s that he wanted to work on energy and the environment. He was drawn to maths and physics at school, so he trained as a scientist. After an early encounter with New Age spirituality, he found his way from Judaism to Buddhism and then to Quakers aged twenty-five. A fascination with psychology and culture led him to work increasingly on the human dimensions of sustainability. Spirituality and sustainability were parallel but separate threads in his life until 2001, when with two other Friends he started the Group on Sustainable Living in Oxford Meeting. This led to setting up the Living Witness project to support Quaker Meetings in developing their sustainability witness. From 2009 to 2013 Laurie lived at the Quaker Community in Bamford, Derbyshire, where he anchored the formation of a new community. He is now back in Oxford Meeting.

Introduction

Quakers in Britain in 2001 issued a 'Call to Action' for the Earth. It spoke of committing ourselves to 'the demanding, costly implications of radically changed lives'. At the time I was a university researcher trying to engage community groups in 'sustainable consumption' projects. With support from the Joseph Rowntree Charitable Trust, I started Living Witness in 2002 as an action inquiry with thirteen Quaker Meetings that had already taken some action on sustainability. I hoped that Quakers might engage readily with sustainable living because of the connection with our established testimony. I also thought Quaker process might offer a way to address the tensions between individuality and community that hamper most experiments in sustainable living.

We held twice-yearly Living Witness gatherings to share and learn from our experiences and support each other. I offered workshops in Local Meetings to explore their leadings and support their witness. Our network grew gradually. Over the years I have led sessions on sustainability with several hundred Quaker groups.

From 2008, the Woodbrooke Quaker Study Centre and Quaker Peace & Social Witness both developed more substantial programmes, and in 2011 Quakers in Britain made the 'Canterbury Commitment' – to become a 'low carbon, sustainable community'.

Some Friends have developed ultra-low impact lives – becoming vegan, giving up flying and driving, reducing their home energy use and installing renewables.

More people have made moderate changes and engaged in some kind of community or political action. But many have yet to begin cutting their carbon footprint. Six years on, we are still finding out what is Quaker about our commitment.

When I set out on this work I was used to writing substantial analytical reports on climate policy and sustainable consumption. I feel I have largely lost that voice. Living Witness trustees have been encouraging me to write a book for some time, and I have struggled to string together more than 600 words at a time. In early 2015 two of the trustees, Andrew Taylor-Browne and Alison Seddon, gave two days of their time to explore with me what might be the core messages from our work. At the same time Ian Kirk-Smith, editor of *the Friend*, invited me to consider writing a series of personal reflections for the magazine.

So, the first thirteen pieces in this book are my attempts to glean insight from fifteen years of working with Friends and I am grateful especially to Alison, Andrew and Ian for enabling me to write them. The remaining articles are a slightly more eclectic selection, some about my personal journey, and some containing reflections that are part of the background to *Gleanings*.

I do not believe our best Quaker sustainability ministry lies in being greener than others, or in shouting louder. We have contributions to make that arise out of the essentials of our corporate faith and practice: being open to transformation by the Light, answering that of God in everyone, and sincerely seeking unity in a way forward. These articles are one attempt to share my experience of those processes.

Laurie Michaelis

Gleanings

The holy mountain

Some years ago I spent three summer days with a group of Young Friends at Pardshaw in the Lake District, exploring their witness to 'that of God in all creation'. In the mornings they generously shared with me – a less-young Friend – their values, visions and ideas for action for a sustainable world. In the afternoons we walked, swam and kept talking.

Their visions were wide ranging: 'we would be in contact with the earth', 'practising arts and crafts', 'we would dance all night', 'no cars', 'no nations or boundaries', 'security, trust, safe in our own homes', 'everything organic, fairly traded, respecting people and earth', 'complement rather than contradict'. I cannot convey the emotional depth and reverence that infused our worship-sharing in the sun; or the effort to wrestle with the complexity of it all, the frustrations, the tensions between our desires and our dreams, our ideals and the 'Spirit of the Age'.

I have experienced many visioning workshops since, with Quaker and other groups. The themes are broadly similar. There is a shared longing for healing in our relationship with nature, with ourselves and with other people. This longing is old and widely shared, at least in faith traditions. In God's holy mountain none shall hurt or destroy and even the lions are vegan (Isaiah 11:6-9). But people disagree deeply on how to get there. They have tried many paths without success.

There is a path up the holy mountain. It might be called 'deep nonviolence'. For me, it is grounded in Quaker testimony and modelled by Mohandas Gandhi and many others. It has strong Buddhist and Taoist resonances. There are three fundamentals.

First, being patterns and examples. If we hope the world will change, we start with a readiness to change ourselves, to learn from the experience and to share it with others. If we see darkness around us, the first step is to let the Light show us our own darkness and bring us to new life.

Second, answering that of God in everyone. We listen, reaching deep for the truth others' words may hold for us, prepared to be challenged, to find we have been mistaken. We embrace difference as well as similarity, recognising others' gifts and seeing how they complement our own. Listening to others, we free them to listen to us and begin to build mutual understanding, compassion and empowerment.

Third, seeking unity in a way forward together. We ask how we are led and submit to God's will – which is revealed when we are heeding the promptings of love and truth in our hearts and answering that of God in each other.

Sometimes the path appears to go nowhere. We feel alone. It seems too slow. We fear we will not arrive in our lifetimes. But stay with it and the mountain comes into being around us.

The path does not take us out of the world: it is living out our faith in the world. It brings new perspectives on agriculture, industry, community, travel, housing and the use of money. It reveals injustice, oppression, violent conflict and ecological neglect as symptoms with a common cause in our illusion of separateness.

Deep nonviolence applies in our families, at work and in politics; by practising it in one sphere, we develop our capacity to practise it in others. We learn how everything is connected.

This is the path that love requires. Occasionally, we may find companions to walk a while with us. That is a profound joy.

Deep nonviolence

In my twenties a friend took me to see *Gandhi*. The film portrayed Mohandas Gandhi as a role model for nonviolence and integrity – principles that shaped his choice of food and clothing, his treatment of people and animals, and his Satyagraha movement for Indian liberation. It made a strong impression on me. I already knew that I wanted to work on energy and the environment; this made the connection with truth, justice, equality and simplicity, with a spiritual path and a way of life.

Gandhi explained 'Satyagraha' as connecting truth with force or firmness. Through noncooperation the movement persuaded the British to let go of India. Climate change and sustainability are different kinds of problems caused by the choices of the vast majority of people. If force is used to solve them, it will take the form of either a minority imposing control or the noncooperation of the planet. Is a deeper nonviolence possible, in which people choose a better path not because they have to, but because it is the collective will?

Adam Curle was a Quaker with extensive experience of mediation in violent conflicts. He was the first professor of Peace Studies at the University of Bradford. In his book *Taming the Hydra* he writes about how violence to people and our planet are perpetrated by the international system of institutions motivated by money and power. He describes the problem as arising not 'out there' but in our own lives and psyches, through what Buddhists call the 'three poisons' of desire, aversion and ignorance.

The root poison is ignorance of our own true, connected nature – the delusion that we are separate beings. This might be the 'ocean of darkness' that George Fox saw.

Quaker discipline and process enables us to experience and act on the truth of our connectedness. In Meeting for Worship it begins with the promptings of love and truth in our hearts, the Light that shows us our darkness, out of which we must grow into new life. We continue to grow through listening deep into each other's ministry and to blossom in the unity of the gathered Meeting.

The fruit is in what happens when we leave Meeting for Worship – which enables us to live better in the world and 'excites our endeavours to mend it'. We are called to make our lives patterns and examples, expressions of Quaker testimony. Our relationships with others are part of this. Answering that of God in everyone is not just a matter of 'seeing the good' in people. It means openness to being transformed by our relationship with them – being answerable to that of God in them.

In our 'endeavours to mend' we are unlikely to succeed by opposing people whom we consider wrong. Deep nonviolence means working for unity. From our experience in Quaker Meetings for Worship for Business we know this often requires long, patient waiting. Averting catastrophic climate change depends on an end to fossil fuel use, globally, within a few decades. It seems there isn't time to build a collective will. But, barring a benevolent world dictator, there is no alternative.

People often ask whether nonviolence works as a strategy. For Gandhi it was an expression of truth and love, and the only way to real peace. The theologian Walter Wink wrote that 'nonviolence and love of enemies… are the only means known for overcoming domination without creating new dominations'.

My experience is that people mostly don't respond well to persuasion. Winning a war or an election, or even an argument, cannot lead to true peace. What works is also what faith requires of us.

Living with difference

When religions engage with sustainability, they often start by revisiting scripture. Christians and Jews may refer to Psalm 24, which asserts that 'the Earth is the Lord's, and the fullness thereof'. Quakers have found much in the writings of early Friends to support a connection with nature and a duty of care and nonviolence towards it.

The next step is an authoritative pronouncement, whether in a papal encyclical or a minute from Britain Yearly Meeting, calling all disciples of the faith to act.

But where do we go after that?

In essence, sustainability is a journey of spiritual transformation to support our flourishing as individuals, communities and ecosystems. It means accepting unpalatable truths, owning the roots of suffering in our way of life without being consumed by blame and guilt. It means awakening to a new sense of self and of our relationship with other people and all life, now and in the future. And it means finding our capacity as agents for change, discerning how best to use our power and energy.

I do not believe there is a 'right' path for this journey. We have different access points and different roles to play; we find our own ways of getting stuck. For some, opportunities to connect with nature may bring new passion and commitment, but many of us are more motivated to act by seeing the human inequalities embodied in our way of life. Some feel energised by connecting with their anger at the powers that be; others are more inspired by a vision of a transformed world, or must get their hands dirty with the quiet, practical work of building a new society.

In her 'Work that Reconnects', American Buddhist Joanna Macy offers practices to address despair and denial that find resonance with many Quakers. The one I have experienced most often is 'Truth Mandala', a ritual sharing of grief, anger, fear and emptiness in response to the state of the world. People often find this a moving and cathartic process, bringing a sense of closeness in a group.

The Light that shows us our own darkness is often unwelcome – associated with blame and guilt. Here I have found another Buddhist, Pema Chödrön, particularly helpful. Her book *Comfortable with Uncertainty* offers spiritual exercises to embrace difficult feelings and nourish compassion for ourselves and others. Our lowest points can be the connection to all humanity; darkness really can be the passage to new life.

An aspect of the journey that many people find hard is to do with holding tensions: between truth and compassion, between prophetic calls for change and gratitude for the blessings of our lives and between individual and collective responsibility. This is where Quaker religious experience comes into its own.

In Meeting for Worship we listen to all contributions, reaching deep for the meaning they hold for us. We may be able to take on board painful truths while also finding the calm equanimity that loves people as they are, the world as it is, perhaps even ourselves as we are. We might find we can live with opposites, accepting them as different perspectives on one reality.

Sometimes we experience a gathered Meeting for Worship – a kind of collective consciousness which can be the bridge to a sense of community with all creation. Then there can be unity in a way forward together, giving expression in our lives and the world to the Quaker testimonies: peace, simplicity, equality, truth and sustainability. As a spiritual community we can value the varied gifts Friends bring and accept that we do not each have to do everything.

Call to adventure

Looking around the circle of Friends in Meeting for Worship, I am conscious of the diversity of our journeys. Most of us are probably in the second half of our lives; we have had jobs, perhaps careers, children, marriages, divorces and bereavements. We may have found a sense of purpose; and we may have lost it. All of us have experienced difficult passages. Some of these are ongoing, evident in frowns and pursed lips.

I find it wonderful how some people emerge from dark and desperate times to become wise pillars of our community, rich in self-knowledge, supporting others and working for a better world. Others feel overwhelmed by the darkness. Yet we can never be sure where our journey will lead until it is over.

In his studies of myth, Joseph Campbell mapped common features in the legendary journeys of figures from Ulysses to Moses, Buddha to Parsifal. The hero begins embedded in conventional life. There is a call to adventure. A threshold is crossed – descending into an underworld, or embarking on a voyage to a distant land. There may be trials, battles with divine and demonic forces, enchantments, injuries and even death. The hero returns transformed, bearing salvation for others and sometimes world healing.

Climate change is the consummate call to adventure, for each of us and for humanity. We in the rich world are recognising the unsustainability of our work-and-spend way of life, material comforts and conveniences.

Unless we make profound changes we stand a good chance of killing off most of our own species as well as the rest of life on Earth. Some Friends have been concerned that our sustainability witness is motivated partly by fear, and so is not Spirit-led. Even if this is true, transformative journeys can begin with selfish motivations, or efforts to preserve normality. On the road the hero finds that the self, the normal, must be abandoned, and is rewarded with unimagined gifts. The journey's true destination is completely unexpected.

Climate change will bring soul-wrenching hardships. Yet the worst fear may be that of our own shadow. Perhaps what humanity has to lose doesn't really mean anything. What emerges may exceed our dreams. There may be a discovery of deeper truth, a more integrated sense of life, of meaning, of purpose, of the world.

Stories of collective transformation are scarce. The Exodus is one example which largely follows Campbell's formula. It is not good news for us. Of the generation of Israelite slaves who escaped Egypt with Moses, only Joshua survived to enter the Promised Land. He had shaken off the slave mentality to become a wise leader and warrior.

So far, humanity has largely refused the call of climate change, remaining enslaved to the materialistic individualism of our time. The adventure doesn't look much fun; the required heroism seems beyond most of us; and the destination is in the distant future. But Campbell writes that the hero who refuses the call becomes a victim, living bereft of meaning in a world that turns into a wasteland.

I often encounter Friends who say it is too late for them to try and live sustainably. They may have journeyed enough; now they are putting their affairs in order. They may have invested all their energy in other moral issues. They may see only desolation ahead. But I believe they still have a contribution to make. Those embarking on the journey can learn from their wisdom. As Canadian sociologist Marshall McLuhan put it, we are all crew on Spaceship Earth. This journey needs everyone, even if only to help us across the threshold.

A simple lifestyle freely chosen?

The British government regularly surveys public attitudes on climate change and energy. Two-thirds of respondents say they are 'very' or 'fairly' concerned about climate change. Four-fifths support the use of renewable energy. They agree that people who fly should bear the cost of the environmental damage, but they do not like the idea of increasing taxes on car users. When it comes to their own behaviour, most people feel they are doing enough for the environment and do not want to change.

Policy researchers talk about the 'attitude-action gap': people say they care about sustainability, but do little about it. This translates to the national and global scale as politicians make grand statements about the urgency of addressing climate change, but cannot bring themselves to implement the policies needed. This shouldn't surprise us. It is hard enough to do things we know are good for us – like eating healthily and exercising. Can we really be expected to change our habits and restrict our appetites in the interests of the future of humanity and life on earth?

In Western society the assumption is that we are responsible for our actions; that we make choices consistent with our attitudes and values. But agency – the ability to form intentions and carry them out – is illusory, at least some of the time. Researchers find that we often become aware of our intentions after our actions, and that we change our attitudes to conform to our behaviour.

For most of my adult life I have experienced inner turmoil around food and eating. Part of me thinks I'm fat and tells me to eat less.

This part connects up with ethical concerns about the violence and inequality of the food system; another part of me craves sweet, fatty food, especially ice cream, which is slightly out-of-bounds for a vegan; and another tries to plan sensible, healthy eating. The result used to be a seesaw of fasting and bingeing.

There are similarities in the way real communities work – and in national politics. Much of our collective behaviour is habitual, following conventions. There is an ongoing turmoil of voices trying to change things, some trying to dominate. Western democracies seesaw between liberalism and control. This is better than the violent chaos in some parts of the world, but we lack the capacity to develop a consistent, collective will for a sustainable society – one that can address the roots of global inequality, violent conflict and climate change.

So, what hope is there for becoming a low carbon, sustainable community – as Quakers, never mind for humanity as a whole? I do think it is possible, but it won't happen by simply making a decision and expecting it to be carried out.

In recent years some balance has emerged in my relationship with food; a kind of interior community of intention, able to make and sustain healthy and ethical choices. It has been a slow process of alignment, accepting and loving each of the parts, finding ways of managing my life so all of their needs can be met. Gradually, they have come to accept each other and stopped trying to dominate each other.

Quaker practice involves 'sitting lightly' to our personal positions and desires to control outcomes as we listen for God's will. In so doing we can usually find a way forward together. Occasionally, we find that sweet spot where we can answer that of God in each other, embrace the gifts arising from our diversity and discover that our individual wills are part of God's will. Letting go of ourselves, we find ourselves and much more.

Not a notion but a way

In the 1990s I was on writing teams for several reports of the Intergovernmental Panel on Climate Change (IPCC). This UN body carries out definitive assessments of climate change, its impacts and what can be done about it. I found it both challenging and highly rewarding, being part of this community seeking with great integrity to map out, distil and communicate the truth.

We hear statistics like 'ninety-eight per cent of scientists believe in climate change'. But good science is not about belief, voting or consensus.

There are parallels between scientific method and Quaker faith and practice as truth-seeking disciplines. The Religious Society of Friends emerged in the same decade in the seventeenth century as the Royal Society, which exists to develop and promote scientific truth. Scientific method centres on a willingness to observe, to question, to experiment and to learn from experience. Scientists' findings must be replicable, and are tested through forensic examination by other scientists.

Our testimony to truth is a central pillar of Quakerism. In Quaker practice, spiritual truth is known experientially, inwardly: 'Take heed, dear Friends, to the promptings of love and truth in your hearts' (*Advices & queries* 1). It is tested through feeling how it sits with love and through careful listening as we reach deep for the meaning in each other's ministry. It is tested alongside other contributions as we seek unity in Meeting for Worship for Business. And, ultimately, it is tested in our lives and actions.

The scientific method builds up knowledge about the objective physical world. It is often said to be morally neutral and, although that is questionable, it develops a map without telling us which path to take. Quaker process is complementary to this, helping us to map our subjective experience, to choose the path and find motivation, company and support as we walk it.

As Friends we are not concerned with beliefs and creeds. The truth we seek is 'not a notion, but a way'. Truthfulness and integrity are part of our testimony but there is much more. Truth is about the Grace of God as shown in our lives and Meetings. In biblical terms, the person of truth walks in God's ways. The word 'truth', like its Hebrew equivalent *emet*, is rooted in meanings to do with commitment, faithfulness and firmness, rather than accuracy.

In 2011 Britain Yearly Meeting (BYM) faced up to the truth of climate change. We committed to become a low carbon, sustainable community, saying: 'We need to arrive at a place in which we all take personal responsibility to make whatever changes we are called to. At the same time, we need to pledge ourselves to corporate action' (BYM 2011, Minute 36).

Friends and Meetings are engaging with this commitment with great energy and seriousness. There are, of course, tensions and doubts. I used to be approached by Friends troubled that others in their Meeting did not 'believe in climate change'. Now there is growing concern that it is too late to act. But I don't want to convert people. What matters is not belief but action. And persuasion rarely works, but living our values can be contagious. Faith means being true to God's path, the path we are called to, even when we have lost hope. If we can walk cheerfully along it, we are likely to attract others. As it says in Psalm 85:10-13:

> *Love and truth meet together; righteousness and peace kiss one another; faithfulness springs from the earth and righteousness looks down from heaven; God also gives prosperity, and our land yields her plenty; righteousness goes before him as he puts his footsteps to the road.*

Transformational listening

When I work with Quaker and other groups on their responses to climate change, people often focus either on what the government should do, or else on actions individuals should take in their lives.

The top-down approach, where powerful people and institutions direct change, is the way issues are mostly discussed in the media. It corresponds to many of our experiences at school, at work, and perhaps in conventional religions. A lot of grassroots activism is directed towards top-down action, advocating changes in government policy. Quaker campaigning often takes this form because we are concerned about things that governments seem to control, like going to war or exporting weapons. But we also have a strong commitment to searching out the 'seeds of war' in our own way of life and 'letting our lives speak'.

Climate change is a different kind of concern. It is obvious that government is not the only player and we need to change our lives. We also need to transform the everyday practices and cultures of industries, communities and just about every human institution. Stopping climate change means abandoning cheap fuels, technologies and ways of life that seem to have worked for us by bringing unprecedented levels of consumption, comfort and convenience.

It takes huge, consistent effort to stop people doing what they want to do. We have just about managed it with tobacco in Britain. In the case of energy and climate change, however, the messages and incentives keep changing. Politicians of all parties have lacked the stomach for the bold measures really needed, especially those to reduce energy consumption. Successive governments have encouraged individuals to 'do our bit'. The latest fashion is 'nudge' policies, trying to shape people's behaviour in subtle ways by making low carbon choices easier and more attractive.

There is another way, which is to treat people as moral agents; entering a dialogue about the kind of world we want to live in and the implications of our choices. Our political system doesn't lend itself to such conversations. And climate change is hard to talk about for all sorts of reasons. Most people can't get their heads around the scientific, technological and economic complexities, never mind engaging with the emotional and existential implications.

Effective climate action needs to involve the whole world. Nobody is in charge, so it must be voluntary. It's not much good everybody thinking they have the answer and trying to change each other. We have to answer that of God in each other – treat each other as subjects, not objects. We have to be answerable to that of God in each other – willing to be changed by the encounter. That probably means people listening to each other – negotiators listening to each other at the climate talks, governments listening to citizens, perhaps even activists listening to politicians.

At a recent academic workshop on 'psychosocial' approaches to sustainability, I asked some participants about their experience of talking to their families about climate change. They just laughed. Our families are often the hardest to engage. But if we can't have such conversations with those closest to us, how can we expect to be part of a wider transformation?

As Quakers we inherit principles and practices for listening ourselves into unity. They work. Yet, when we talk about climate change, we often get into the same arguments that are going on everywhere. We should use our listening practices more amongst ourselves. And we should find more ways of sharing them with others.

Being a transformational community

I am a natural hermit. Of the last thirty-five years, I've shared a home for only twelve. I thrive on solitude and have developed a peculiar low-input lifestyle that few people want to share. I'm also part of a wider trend towards single-person households. There are many factors – divorce, longer life expectancy, technology and services that make it easier to live alone. Mostly, this causes higher resource use because we all have our own fridges and central heating.

In the 1990s my research on sustainable consumption brought home to me that community is a key to sustainability, not just because it's better to share things. People live and behave in ways that are normal for their social circle. In the industrialised world today that means driving, eating meat, keeping homes at 21°C and flying abroad for holidays. If we are going to change our way of life, we'll need to do it together.

So, I had to be part of a community. For me, that took the form of Oxford Quaker Meeting, and then the wider Living Witness network of Friends and Meetings developing Quaker approaches to sustainability. For a while I lived at the Quaker Community in Bamford – although I was able to choose secluded, hermit-friendly accommodation.

Quakerism suits individualists. With no creed and little teaching, some attenders mistake it for 'anything goes'. Voices can be heard, including the small, shy ones, including the angry and the conflicted.

There is the potential for the Spirit to break through and change our sense of what is normal. But we also have a discipline of letting go of our positions as we seek unity in a way forward.

Sometimes we try to reach unity by looking for commonalities. I have found more reality, and more usefulness, in embracing difference. We come to community with different gifts, different wounds, different needs and different assumptions about the purpose and meaning of our lives. This makes for both creativity and conflict. Perhaps the hardest thing for any community to do is to stay real – open to each other's truth and pain, open to change, on the edge of chaos.

Community can be a place of personal transformation. More importantly, it can be an incubator, the place where we develop the capacity to work for change in the world. It is so easy to see war, injustice and ecological degradation as resulting from people being bad, stupid, or in the wrong position. But as Parker J Palmer put it: 'Community is that place where the person you least want to live with always lives.' Getting rid of people is not an option. In community, we can learn to tender our anger into love.

One of the greatest gifts I have received in community – though it feels unwelcome at the time – is the opportunity to see myself through others' eyes. Conflict can be a mirror as we try to understand what in us is causing or contributing to the problem. Working through it takes time and work, learning to listen, to hear and understand others, to recognise their needs and their ways of expressing them, which may be so different from our own. It means being sensitive to their feelings. It means trusting and empowering them, taking what they want seriously, being willing to support them even when their priorities are different from our own.

We can create community in all sorts of places. Sustainability is about developing a true community of all life. Our Quaker Meetings could be a good place to start.

People and place

Friends often assume that I grow my own vegetables. It is part of a green stereotype, perhaps influenced by the 1970s sitcom *The Good Life*. That stereotype has some validity. The Transition Movement has a focus on local self-reliance. This is partly about reconnecting people with soil and community and partly about resilience, as we face growing uncertainty about climate, economy and government.

I'm definitely not a gardener, although I like to be on friendly terms with the person who does grow my vegetables, and to be able to cycle to his farm near Oxford for my weekly kale supply. For the record, he's called George.

The urge to reconnect is evident in politics. The long-running polarity of left and right, equality and liberty, is giving way to a 'politics of belonging'. Nationalist parties want to revive the association of people with place to protect identities, cultures, lifestyles and livelihoods. In green politics the concern is more about developing connections across traditional boundaries – between different groups of people, and between humans and nonhumans.

So, the Transition Model of community building aspires to inclusivity: 'Everyone is needed.' A resilient community can welcome migrants because its members feel secure. But can this be sustained as local infrastructure and services crumble, with hundreds of millions on the move globally? Ownership and connection so easily leads to exclusivity, conflict and violence.

As a Jew, I have some ambivalence about localism. I feel at home here as a fourth-generation British immigrant but my roots are shallow and urban. My heritage is one of being on the edge and on the move. In the Torah, the land has some of the qualities of a person. It is an agent of God: 'Keep... my laws... lest the land... spew you out' (Leviticus 20.22). We cannot fully own it because 'the land is mine', says the Lord (Leviticus 25.23). Like us, it requires rest and restoration – its sabbatical years and its jubilees. We are enjoined to welcome others to share it with us: 'Love the stranger... for you were strangers in Egypt' (Leviticus 19.34).

In Quaker terms, what does it mean to be answerable to 'that of God in all creation'? As in human relationships, I think it means cultivating respect, understanding and caring for the land. It also means being willing to be transformed by my relationship with it – to learn from it, and to live within its carrying capacity.

Healthy connection to the land brings a sense of commitment rather than entitlement. Its requirements include simple living and a willingness to share – including openness to others who may have different habits and lifestyles. They may also have different ideas about what to do with the land. At present, we haven't got it right as a nation. The United Kingdom's arable produce would sustain the population twice over, but we import half of our food. We feed more of our crops to animals than we eat ourselves. Meanwhile our uplands are – as George Monbiot puts it – 'sheepwrecked', shorn of the prehistoric forests that could have helped manage our watersheds and provided some of our energy needs.

We come back to familiar spiritual and ethical tensions, trying to walk the middle way. Can we be connected without being attached? Can we be open to the inevitability of change – knowing where we are, where we have come from and having a sense that we may have to move on? And can we cope with the ultimate challenge of corporate discernment, finding a way forward together while respecting our own needs, those of the other inhabitants of the planet, and those of the land?

Losing liberty?

And you will know the truth, and the truth will set you free.

John 8:32

A simple lifestyle freely chosen is a source of strength

Advices & queries 41

My lifestyle choices – especially cycling and being vegan – are partly an assertion of my freedom, self-sufficiency and individuality. It helps that they are also cheap and low-carbon. But for many people sustainable living suggests constraint. It challenges precious freedoms – to drive, fly, eat meat, run the central heating at full blast. It cuts against the spirit of our age, which values individual freedom 'in the pursuit of happiness'. In the culture of consumerism, happiness is mostly pursued through buying stuff and through personal comfort and travel.

Sociologist Zygmunt Bauman has written about the inevitable tension in community between freedom and security. We give up freedoms in order to belong and to meet the requirements of fairness and social harmony. Even in the most libertarian societies, personal freedom may be constrained where it harms others, although that is a difficult negotiation, and the rich and powerful come off best. In the French Republican slogan, *liberté* and *egalité* are only reconciled by *fraternité* – by caring about each other and about society.

It is increasingly obvious that climate change is causing harm. It will wreck billions of human lives as well as destroying species and ecosystems.

At some point our freedom to burn fossil fuels must be constrained. If the Paris Agreement of 2015 is to have any meaning, that moment must come soon. Could it be a joyous moment?

Quakerism can appear a libertarian religion. Many people find a home with us because of the freedom of belief, and of our interior practice in Meeting for Worship. In our Local Meetings we want to be inclusive, rather than emphasise the discipline that underpins Quaker faith and practice. Phrases like 'Gospel Order' are hard to explain, especially when few seasoned Friends are really sure what they mean. At the first encounter, 'Right Ordering' can seem slow and stifling. Familiarity comes gradually, through deeper involvement in Business Meetings, in Yearly Meeting and national committees.

In western society, freedom comes to be identified with unlimited choice. Nobody tells us what not to do. We can put anything we want in our shopping trolley. But research shows that, even when shoppers think they are making free choices, what they actually put in their trolley conforms to the norms of their social group.

There is another kind of freedom that comes from the acceptance of limitations, celebration of our world, loving the age we live in, even as we are led to work for change. It is when we are attuned to our context – to our community, to God – that we have the strongest sense of agency, of being part of a larger agency, and so of freedom. It may be the freedom of going with the flow, or the freedom of opposing the 'domination system', supported by our faith.

This kind of freedom – choosing a structure, a discipline to live in – can still look to others like a loss of liberty. But constraint is only a loss of freedom when we do not choose it.

Through consistent spiritual practice, listening and watching, inwardly and outwardly, we may awaken to our own shadow, which may have been controlling us; we may awaken to the leadings of God, whether revealed in our own hearts, in the ministry of others, or in the world around us; and we may awaken our capacity to follow those leadings. This is true freedom.

Making a connection

I meet a lot of Friends who want to know how they can communicate better about climate change. Some are frustrated or distressed because they think too little is happening to implement our Quaker commitment to sustainability. Perhaps others in their Meeting do not share their concern, their clarity about the action needed or their willingness to measure their carbon footprints, change their own lives and speak out in the world. People do not seem to be listening.

Although Quaker writing and ministry on sustainability goes back a long way, we are only beginning to acknowledge it as testimony, as a core part of being Quaker, and to understand what that means. Friends are still trying to convert and persuade each other. Often, they use arguments and language from the mainstream media. Some Friends remain uncomfortable with the idea that there might be any distinctive Quaker approach. I certainly doubt that there is a particular Quaker message.

A large part of conventional religious ministry is to do with instruction. The minister is supposed to be God's servant speaking to the congregation. Quaker ministry in Meeting for Worship can sometimes feel a bit like this, focused on getting a message across.

A painter, musician or poet may have a message in mind, but if the art is to live, the artist must let it go. If it is any good, it will draw people into relationship with it, to make meanings and discoveries that the artist never imagined.

Like our Quaker sustainability ministry, much climate change art is rather earnest and focused on changing the audience.

In thirty years of writing and speaking about sustainability my messages have rarely, if ever, been received as I intended. People filter words through their own experience. When I give a talk or a workshop people may remember that I arrived by bicycle, but not what I said. They might remember one tangential bit of information, but not my core theme. Yet, once in a while, I meet someone who tells me how they changed their life after one of these events.

At its best, ministry is not the transmission of a message: it is an answering of that of God in everyone. It is a making of connection. It enables each of us, speaker and listener, to contact our own experience of the Spirit, of love, truth and faith. In Meeting for Worship it may open up a space, drawing us into unity.

This opening of a channel for the Spirit, in Meeting for Worship or in the world, can take many forms. Some of the most powerful ministry starts with the effort to change ourselves. It might be shared through just being. We may share our experience in words, but actions speak louder, and the sharing might sometimes reach deeper if it uses visual media, music, dance or drama.

Our Quaker sustainability ministry is bubbling up, in Britain and elsewhere. Friends are painting pictures and telling stories about climate change. They are changing their lives and participating in creative direct actions. They are working in politics and in their local communities. There is an ongoing task of nurturing this ministry, in our Local Meetings and in the Yearly Meeting.

I still encounter the assumption that Friends without environmental expertise have nothing to contribute. But we need the ministry of Friends experienced in eldership, clerking, nominations, community building, working with children and conflict transformation. Perhaps the most powerful Quaker ministry is that of listening, of our own willingness to be transformed by the encounter with God in the other.

Higher power

Living simply can be anything but simple. At the Britain Yearly Meeting Sustainability Gathering in 2016, Friends spoke of their dilemmas in cultivating low carbon lives; air travel to visit grandchildren is just one example.

Many of our 'needs' are modern inventions. Until the 1960s most travel in Britain was by bicycle or public transport, and in the early 1970s the average indoor winter temperature was 13°C. Yet few in our society would seriously consider doing without cars or central heating, even knowing that this new technology threatens the survival of our species. Is this a kind of mass addiction? Is it rational self-interest? Or is it just the new normal?

I am wary of claims for rationality. In major life choices, even when I have thought through the options carefully, I realise looking back how I was swayed by my assumptions and habits of thought, by the opinions of those around me and by motivations I did not fully acknowledge at the time.

Day-to-day choices are even messier. I have a complex relationship with food. Sugar addiction seems to be at least part of this; so is coping with social norms and pressures. Consuming a little sugar results in cravings for more and can lead to a binge; but controlling my eating too tightly can also be a trigger. There is a mixture of feeling something else taking over, of self-disgust and self-justification. It is not a perfect analogue with our society's fossil fuel addiction, but there are common elements.

My personal journey of integration involved letting go of the effort at self-control, recognising that no one part of me can be in charge. At the same time it also involved monitoring my eating, as well as watching my feelings and thoughts carefully and sitting with them. There was an important step of telling myself that 'God loves me'. Eating stopped being about virtue and guilt, and became a matter for discernment – sensing what's right for my body, feelings and ethics.

In twelve-step programmes, the first step is acknowledging that we are powerless over our addiction. The second is to 'come to believe that a power greater than ourselves could restore us to sanity'. Similarly, George Fox advises:

> ...*whatever ye are addicted to, the tempter will come in that thing ...Stand still in that which is pure, after ye see yourselves; and then mercy comes in... do not think, but submit... Stand still in that which shows and discovers; and then doth strength immediately come. And stand still in the Light, and submit to it, and the other will be hushed and gone; and then content comes.*
>
> *Quaker faith & practice* 20.42

Applied to our challenge of low carbon living, there are some clear messages. We can start by being willing to see what we are doing to ourselves and other life on our planet. That means recognising the violence inherent in our normal behaviours such as driving, flying, using gas and electricity, and eating meat and dairy products. But we have to stop squirming, let go of the focus on blame, guilt or virtue, stand still in the Light and allow ourselves to be guided.

As with addiction, mutual support is important. We can share experiences as we hold ourselves to account without judgement or excuses. We can recognise and remove temptation in our lives together. We can witness without condemning or condoning each other's carbon-intensive choices. We can celebrate small successes. Through all this we gradually establish a new, low carbon normal within our own community. And that strengthens our capacity to work for change in the society around us.

Seeking unity

I have been at many different kinds of gathering where, after some days together, we felt a sublime connection and a capacity to change the world. It happens in a particular form at Yearly Meeting: the Light shines, a way opens and some intractable conflict is suddenly resolved. Lis Burch recalls that, as Yearly Meeting clerk in 2011 at Canterbury, her notes before the Friday morning session read: 'Thinking is still diverse… no obvious corporate commitment.' But she says we were 'spiritually seized by the scruff of the neck' that morning and made our commitment to become a low carbon, sustainable community.

Climate negotiators in Paris in December 2015 may have felt similarly propelled. Like our Quaker commitment, the Paris Agreement is aspirational. The path is not easy. In order to stop climate change we must move from good intentions to changed lifestyles, technologies, industries and human relationships. A Transition Movement slogan puts it clearly: 'Everyone is needed.' Perhaps not quite everyone: we need a critical mass of people acting together worldwide. But humanity today seems more concerned with difference, with doubts about making common cause with others. How can we unite across the divides between religions, regions, ethnicities and nations?

The urge to differentiate is not just selfishness – it is an expression of a need for belonging, for real relationship with others 'like us'. Communities have atrophied in our individualising, globalising world. Rebuilding local and national identity might seem to be a way of rekindling connection. But people perhaps forget how hard it is to reconcile individual freedom and creativity with the security and 'belongingness' of community. Part of what excites me about Quaker faith and practice is that it sometimes manages this balancing act and generates something more. It does so through a discipline that includes:

- Being open to change, letting the Light show us our darkness and bring us to new life.
- Answering that of God in each other; reaching deep for the truth their words may hold for us.
- Setting aside our own positions and expectations as we seek a way forward together.

This spiritual discipline differentiates Quaker unity from the consensus methods used in the Transition and Occupy movements, although both were influenced by Quakers. Participating in secular or interfaith meetings usually reminds me why I am in love with Quaker process: allowing ministry to emerge out of stillness; careful listening, with time to absorb each contribution; striving to discern what is right to be said, rather than speaking out of ego or defending a position; and the servant-leadership of the clerk.

I love the way, at Yearly Meeting, there is time in opening and closing worship for ministry that doesn't fit the agenda. I love the way all perspectives can be shared, and we can find unity in a vision and a course of action. This discipline is not easy. The Spirit of the Age is alive in all of us. I have often experienced conflict in Quaker Meetings where differences in personality or personal history turned someone into an enemy. We so easily slip into habits of self-defence, judgement and blame. Yet our Local Meetings for Worship and Meetings for Worship for Business are a constant opportunity to practise, to correct our mistakes and hone what Buddhists call 'skilfulness'. If we do so, this is a religion that enables us to live well in the world and nourishes our efforts to mend it, answering that of God in every one, all creatures and all creation.

Other pieces

Answering that of God in everyone

Six years on from our Canterbury Commitment to 'make whatever changes we are called to', many Friends have apparently not yet been called. Perhaps we are not so different from the rest of humanity or the people in power. There is frustration among Friends working for change.

Some feel the solutions are obvious. Often there is an expectation that if people only knew the truth – the science of climate change, the emerging impacts on ecosystems and vulnerable people – they would respond. Some are more focused on taking action and addressing power dynamics, seeing the problem as lying in 'corporate greed' and vested interests. Others believe that Quakers need a new spirituality of connection to nature and all life.

All of these approaches are needed, but none is sufficient alone. In trying to connect with others in responding to climate change (or perhaps any issue), there is a tendency to focus on an approach that may be our own area of strength, or where we've had our own breakthroughs. We might achieve more by being sensitive and responsive to those we are trying to engage.

Asked what Quakers believe, a common response is 'that there is that of God in everyone'. In fact George Fox talked about 'answering that of God in everyone'. There was no question for him that it is there; the question was, and is, how we respond to it.

At its heart, Quaker practice is based in listening both inwardly and to others. We are advised to reach deep for the meaning in each other's words, to seek the truth their opinions may contain for us, to think it possible we might be mistaken. Answering that of God in the other means also being answerable to it, ready to be transformed by it.

Another favourite Quaker phrase is 'speaking truth to power'. Andrew Clarke, former general secretary of Quaker Peace and Service, suggested that it might be better to 'seek truth with power'.

There is a close connection with Martin Buber's message about entering an 'I-Thou' relationship with the other, human or not, recognising our co-subjectivity. When we recognise others as subjects, beings like ourselves with their own connection to God, we may begin to let go of the idea that we can or should change them. They have their own journey of transformation to make, and they need to find their own footing. Perhaps it is we who need to change.

Certainly, I have never found it helpful to try and persuade people to do things, although that may just be because I am not very good at it. The best I can do is to make my own choices, make them visible and, if people ask, explain my reasons. Those reasons are not all altruistic. I have many selfish motivations for being vegan and cycling. But sometimes explaining my peculiar journey to people prompts them to think about their own. Nor have I found anger a useful basis for communication or action. Mostly, people respond by putting up barriers. On very rare occasions, in a relationship where the loving foundation is clear and solid, anger can give the impetus to talk about something that needs addressing.

It is in intimate and community relationships that I have found the toughest challenges and have learned most. People have questioned my motivations, self-image and worldview. I have made a fool of myself and experienced shame and regret. But with time and work to answer that of God in the other, I have also experienced the emergence of mutual love, understanding and trust out of conflict.

Joining the movement for a new society?

The world is at a turning point. The 'Spirit of the Age', described by Jonathan Dale in his 1996 Swarthmore Lecture, is on the defensive. That spirit, the culture of competitive, individualist materialism, has dominated Western civilisation for decades during which traditional institutions and the sense of local community have declined. There has been a surge in politics based on the reassertion of national and regional identities, with huge uncertainties about the future balance of power within and between countries.

In this context Quakers may feel that we are on one side of the debate – more comfortable with the liberal, pluralist policies that are under attack than with the identity politics that is on the ascendant. But neither side offers a viable agenda for a transition to a sustainable society. That would entail a transformation in the way we behave towards each other and the planet we live on.

There are many groups working for such a transformation, often with a sense of certainty that we lack. When Friends engage with sustainability, there is sometimes a feeling of burden and guilt. Change is seen as costly, although we tell ourselves that it can also be joyful.

Our Quaker progress has felt slow, despite successive Yearly Meeting minutes calling for a spirituality of connection with the Earth, for transformed lives and for speaking out personally and corporately. We look at the public pronouncements of other faith groups and find ourselves wanting. But if we are finding this journey challenging, perhaps we can have all the more compassion and understanding for the difficulties faced by others who seem to be resisting change. Perhaps we can recognise our agony as the world's agony.

A real movement for a sustainable society would be inclusive. Low carbon, sustainable living would become normal and justified in a wide diversity of groups and cultures. For some the motivation might be compassion for all life; for others it might be to do with religion, national pride or economic rationality.

Friends will find different ways into our sustainability commitment and that is a source of strength. For some the starting point is new knowledge, insight or recognising themselves as part of a meaningful story. For some it is a sense of connection, belonging to a transformational community, compassion for those suffering the impacts of climate change, or unity with nature. And for some it is the experience of their own agency, changing their own lives or working with others for system change.

Whichever path we follow, I believe there is a powerful ministry that we can offer in the wider movement based on who we are already, on being true to our Quaker faith. At the centre of this are the disciplines of openness to transformation, answering that of God in everyone and seeking unity in a way forward. It may be that none of these disciplines are uniquely Quaker – they have similarities in other faith and secular traditions. However, together they form a particular combination in Quaker faith and practice. We could start by fully practising these disciplines in engaging with sustainability in our lives and Quaker Meetings.

Seeking unity
in a way forward together

Some Friends talk about our business method as a peculiarity that slows us down in making important decisions. But our corporate discernment practice could be the most distinctive and most needed gift that Quakers can offer the world. Friends have already influenced activist movements for peace, justice and sustainability, which have generally adopted consensus decision making. A key channel for that influence was the Movement for a New Society, which developed training and handbooks for nonviolent activism in the United States, originally in response to the Vietnam war.

However, activist networks tend to be made up of people with largely shared values and purposes. Quaker method differs from what is commonly understood as consensus in that it is not about participants agreeing with each other; nor is it a process of negotiation between positions. It involves a spiritual discipline of self-forgetting, letting go of expectations, listening deeply to each other and the Spirit, and sincerely seeking unity in recognising the right way forward.

In our sustainability work it may be tempting to think that, if only people knew what we know, and felt the way we felt, they would do the right thing. It may be tempting to join with like-minded others. But achieving a low carbon, sustainable civilisation requires a lot of people to be involved, and they are not all going to adopt our worldview.

As Quakers, we know that it is possible to experience unity and find a way forward together without shared beliefs. We also know something about the discipline required. Quaker processes work well when we attend to the foundations, 'getting to know one another in the things that are eternal', listening to people with differing views and making sure everybody's needs are taken care of. Business Meetings don't work when we try to rush a decision or prejudge the outcome – we may produce a minute, but afterwards Friends may not feel full ownership of it.

On difficult and complex questions, developing unity can take several years. Our sustainability concern is a case in point. It requires ongoing conversations and quiet listening in Local and Area Meetings, and in national committees and gatherings, to develop a shared understanding. We are not there yet.

When we offer Quaker process to others, the results can be amazing. The Quaker United Nations Office has been doing exemplary work at the UN climate negotiations. In the three years leading up to the 2015 Paris Agreement, QUNO held dinners to which delegates from all the major negotiating blocs were invited. They were invited to listen to each other responding to questions that are not normally asked in the negotiations, for instance about their personal hopes and fears, how climate change would affect their home countries and their feelings about the process. Delegates were clear that this process contributed to an improvement in relationships.

I wonder whether Friends could develop a similar ministry in Britain – for instance using our Meeting houses to host community conversations. We could aim to bring together people who do not usually talk to each other and to hold quiet conversations of kinds that they do not usually have – focusing on personal values, feelings, hopes and fears. Perhaps we could address issues like housing, health and transport that concern people of all cultures, ages and political persuasions. We could even surprise ourselves with the new friends we might find.

Darkness and new life

Fifteen years ago I was approaching forty. It seemed like time to live my values more authentically. From my late teens I had been committed to working for sustainability. In my mid-twenties I had become a Quaker. Now, perhaps, these two threads could come together.

The following years were the most turbulent in my adult life – perhaps a classic mid-life crisis. I thought I had made some poor choices by ignoring gut feelings. Now I acted more on instincts, and conflicts kept cropping up. I made messy, hurtful and rather public ends to my career and my second Quaker marriage. I was angry and defensive, blaming situations and other people.

I was encouraged to resign from my Quaker roles and to attend another Meeting. But I insisted on staying in my Meeting – I wanted to work through the hurt and what seemed to me to be a misunderstanding.

At forty-six, single and thinking my work had reached a dead end, I felt I was turned to ashes. There was an ache which I coped with by spending hours every day on long walks. Eventually, I realised that everybody involved in those messy situations had been doing their best – including me. One or two Friends continued to remind me what a fool I'd been but nominations committee came to feel I was ready to serve again. I was appointed as an elder.

There has been a lot of media discussion recently about 'shaming'. Shaming is something that is done to people, which scapegoats them, driving them out of the community. For me, shame was part of a positive process, realising that something in me was the problem, causing conflicts and hurting people. *Advices & queries* 1, with its words about 'the leadings of God whose light shows us our darkness', speaks powerfully to me.

'Darkness' is not badness. It is normal and inevitable that we have areas of partial or distorted awareness. My most shadowy areas are to do with instinct – those gut feelings that got me and others around me into trouble. My learning path was, and still is, partly about discernment. That means listening to my gut as well as to my head and heart, and to other people. It means heeding the signs, especially when they warn against action; waiting for the Way to open. Another part of my path is about understanding people whose areas of awareness and shadow are different from mine, recognising how we misinterpret each other. Perhaps the hardest part is doing the work it takes to truly answer that of God in each other.

My shame shapes my perspective on the challenges of our age – from climate change to violent conflict, injustice and oppression. I am wary of 'righteous' anger and find myself with a particular concern for answering that of God in the perpetrators. Much of the violence is symptomatic of old, deep wounds, and of a system in which we are all complicit – even if we try to distance ourselves by fossil fuel divestment or avoiding products from illegal Israeli settlements.

Accepting my complicity – my shame – helps me recognise my connectedness to everyone, to empathise with others who are in a difficult place on their journey, or have not even recognised their own darkness. Shame can be a source of humility – I would not pretend to know the answers for you or anyone. But it also gives me hope for the world, because knowing I am part of the problem means I can be part of the solution.

Back to the land?

In Yearly Meeting 2003 we began, tentatively, to sound out the depths of our feelings about the situation in Palestine/Israel. For the many Jewish Quakers present the emotions were complex and painful, yet I think there was a sense of heavenly relief in having our shadow laid before the Meeting, and in the recognition of our companionship as Jews. Stevie Krayer's introduction spoke for me, and I think for others present.

In the corridors afterwards, several Friends felt we had paid insufficient attention to 'the issue of land'. There had been ministry in the session about the crimes of Europeans in violating the relationship with the land of the original inhabitants of Australia and North America. We were reminded of the strangeness to them of the Western concept of land ownership. But the issue was not reflected in the minute.

I'm glad about that, because I think the question of our relationship to the land could take up at least a whole Yearly Meeting session, and we would need thorough preparation. It is hard to realise how much that relationship varies from culture to culture, and also within cultures.

Christianity, Islam and Judaism understand the Earth as God's creation, over which humans have a duty of stewardship. In contrast, cultures that emphasise the immanent spirit(s) in all life and non-life often see humans in an equal relationship with nature. But within the monotheistic traditions there are many different perspectives.

Jewish attitudes are complex and diverse. The Torah emphasises that the Earth belongs to God; the corners of fields are to provide food for the poor; the land must have a Sabbath every seven years; and all land purchases are to be revoked in the Jubilee every fifty years.

Yet God gives the land of Canaan to Abraham's seed. The prophets warn that through Israel's misdemeanours the Covenant is forfeit, but they also look forward to its restoration. Over the last 2,000 years, Jews have repeated to themselves a liturgy that affirms the permanence of the Covenant and the aspiration of a return to Zion, to Jerusalem, to the sacred soil promised to Abraham.

But when does a relationship with the land expire? Is the Jewish 4,000-year-old lease simply obsolete? Are we capable of a healthy relationship with any soil after 2,000 years of being moved on, prevented from owning land? Many, if not most, Jews feel a strong attachment to Jerusalem. 'If I do not remember Jerusalem, may my right hand wither away, may my tongue cleave to my palate.' The image of the landscape of Judaea with the walls of the Old City of Jerusalem glows in my mind's eye. Jerusalem the golden. Our 'true home'. It is not clear, though, that being there demands sovereign ownership. Even in Biblical times Jerusalem was shared with its original inhabitants, the Jebusites.

The early Zionists were idealistic. They were making the desert bloom; bringing life to a place of desolation. As bourgeois Europeans, products of the imperial age, they also thought they were bringing civilisation to the existing inhabitants. More apparent now is the land-grabbing approach of the State of Israel. This is often justified by Israeli settlers in terms of the Old Testament Covenant, but it is probably more about military strategy. Israelis talk of the impossibility of defending their country with its original 1948 borders and the narrow coastal strip.

The current approach is grounded in a fear of the rest of the world. Many Israelis believe that the world hates them because they are Jews and that the only response is to establish a fortress.

Some, émigrés from Russia or Middle Eastern countries, have recent personal experience of rejection or persecution. Others are the children or grandchildren of European Jews murdered or displaced in the last century. Israelis must prove that they will fight back at any new attempt by the Arabs to exterminate them, to 'drive them into the sea'. Fear often seems to blind Israel to the treasured land it is defending. It has pursued development at a very high cost, with overexploitation of resources and rapidly rising levels of pollution.

In my own family I experienced two perspectives on the land. My mother's parents were pillars of the Zionist movement. They saw the Israeli stronghold as the key to Jewish survival. They owned a home in Jerusalem, although their primary home was always their flat in London. My father's parents had little interest in Israel. They felt that we should simply be grateful to live in a country where we were more or less tolerated. Their relationship to the Earth was concrete and local to their home in Golders Green, where they were dedicated gardeners.

Is an attachment to the land good for some people, and bad for others? What if two groups of people are attached to the same plot? Is there enough to go around in a world with a population heading for nine billion?

For me, the survival of my culture has always seemed less pressing than the survival of the human species. So, I have spent my adult years working for environmental sustainability. But many environmentalists argue that we need to restore our relationship with the land, pointing to the examples of First Americans and Australians. They blame our current ecological malaise on Enlightenment thinkers such as John Locke with his justifications for land enclosure. Locke drew on the Bible in arguing for the development of land in North America. Perhaps he provides the prototype of the Zionist story about making the desert bloom.

I don't think I have the answers. I think we need to reflect on questions about ownership, about boundaries, about sovereignty.

Always, always, we must look for ways to move forward that do not involve violation of others' sense of their needs and rights.

It was clear from an early stage that the European project of a Jewish state in Palestine was going to violate the needs of the local population. But European and American Christians in the 1940s were not prepared to share their own land with the Jews.

The world has not changed much. As the numbers of displaced people escalate globally, there are many more conflicts looming like that in Israel/Palestine. We must find a new model of nationhood and of our place in creation. And land is a key part of that model.

Living in the End Times?

Armageddon is in vogue. Perhaps there's nothing new in that but the signs now seem particularly ominous. Some years ago I was invited to be on a panel with a Hindu and a Muslim on an Iranian television network. We were asked whether recent floods and other extreme weather events were signs of the End Times as predicted in our various scriptures. It seemed a good opportunity to learn something.

Early Friends saw signs of the End Times in the tumultuous events of seventeenth century England and Europe. Although that was a particularly turbulent period, self-made prophets always seem able to convince themselves that Armageddon is imminent. This is how many Christian movements have started. They usually denounce the mainstream establishment of their time as fallen or evil, claiming that their own members will inhabit the New Jerusalem, fulfilling the predictions of scripture.

The end hasn't happened yet and sects have several options when their prophecies fail: to acknowledge that they were wrong and disband or self-destruct; to reschedule the end, either to a specific date or to the distant and indefinite future; or to reinterpret scripture as essentially figurative.

Liberal Friends have largely adopted the last approach. Eternity is now and we are called to live in the Kingdom today. We have quietly discarded George Fox's habit of ranting against the immoral world, seeking to bring our own darkness into the Light and to recognise that of God in others.

How does this relate to climate change? For many people, environmentalism seems like just another new religion. The warnings of disaster fit the apocalyptic mould. As usual we are being called to wake up and change our lives. We've heard it all before.

For some climate change believers, it probably is a religion. But it is also science. The vast weight of the evidence now affirms that climate change is happening and that there is a substantial likelihood of very nasty impacts.

On our television panel, it was clear that extreme weather is central in the apocalypse stories of all our religions. But we did not think it helpful to make that connection with climate change. We agreed that the real value of these prophetic visions is not in concretely mapping out the end of the world, but in their call to live a moral life. In particular we need a shift away from competitive consumerism – whether to avert climate change or because of our core spiritual values.

End Time movements have sometimes succeeded in their calls to moral reform. Bible-based apocalyptic thinking in the seventeenth century helped to reform European society, ushering the values of the Enlightenment. Could climate change be the stimulus for a different kind of moral renewal? If so, what shape might it take?

Our individualised, materialistic society has been in many ways a reaction against the restrictions of traditional community. But consumerism does not meet many people's needs for community, moral certainty and a sense of meaning. As the Muslim on our television panel rather forcefully reminded us, there is now a tide of converts to Islam.

The fearfulness of this age demands that we as Quakers get clearer about our own gifts – which are not so much about about certainty and a prescribed way of living as a discipline of openness – and offer them to the world.

Recognising our own darkness and answering that of God in other people, we may begin to be able to make the changes needed.

Loving the Spirit of the Age

Quakers have often sided with the 'revolution'. In the 1650s the first Friends were part of the move to turn English society upside down, abolishing old feudal power structures and inequalities. But did Friends help to build the power structures and inequalities of today?

Nowadays Quakers know where their sympathies lie. They have a strong presence not just on anti-war marches, but also in the anti-globalisation movement and at Climate Camp. You don't find many Friends working for large corporations. They are far more likely to be teachers or social workers.

Quakerism is an engaged faith, connecting spiritual practice with action in the world. Those seventeenth century Friends believed the Kingdom of God was coming, literally, in their lifetimes. They were working for real, immediate change.

Recent movements like Occupy and Climate Camp may have had a similar feeling. Their revolutionary movements felt prophetic both in their message to the world that we must change the way we live and organise our society, and in their sense of moral community at camps and gatherings. Decisions were made by consensus. People cooked for each other, shared, provided mutual support, learned and grew together. And like early Friends, they had a sense of being embattled by the powers that be – nonviolent gatherings surrounded by police in riot gear.

Yet something jars here.

The organisers of Climate Camp said our ecological and energy crisis 'has its roots in a capitalist system'. But if you go back to the emergence of that system in the eighteenth and nineteenth centuries, you find rather a lot of Quakers.

Much Quaker work now relies on chocolate money – charitable funds left by the Rowntrees and Cadburys. Earlier Quaker capitalists included the Lloyds, ironmakers turned bankers, and the Barclays, whose bank emerged from a goldsmith's business.

Quakers helped shape the Industrial Revolution and the financial system that funded it. They thrived in that system. Hard work and plain living meant that they reinvested the money they made. Their strong national community brought mutual support, scrutiny of each other's business practices, exchange of skills and an ability to innovate. And the public perception of Quaker integrity made them a trusted 'brand'.

Capitalism emerged as a highly moral system. Quaker and many other industrialists used their wealth and power to improve their employees' quality of life. But there was a contradiction. Quakers were producing the goods and providing the finance that underpinned the new consumer society – the Spirit of the Age that they condemned in their Yearly Meeting minutes and epistles.

Speaking to the world gathering of Friends in 2007, Marion McNaughton quoted the Benedictine, Jean Leclercq: 'We must love the age we live in.' Can we love the competitive, materialistic individualism of the Spirit of the Age? It underpins much that is unsustainable in our society from climate change and oil wars to the over-reliance on consumer credit.

A growing number of Friends are questioning the Western way of life – finding ways to live without cars or air travel, turning down the central heating and cutting their consumption of meat and dairy products. It is harder to question individualism itself. Quakers are perhaps the most individualistic of faith groups, many being attracted from other religions by the freedom to 'decide for yourself' in the non-creedal silence.

Yet, the Quaker way combines individualism with a strong sense of community. Maybe that is the shift that is needed in the Spirit of the Age.